BEVERLY HILLS CARS

BEVERLY HILLS CARS
LUXURY AND POWER IN THE CITY OF FANTASIES

COLIN BURNHAM

Published in 1991 by Osprey Publishing,
59 Grosvenor Street, London W1X 9DA

© Colin Burnham 1991

All rights reserved, Apart from any fair dealing for the purpose of private study, research, criticism or review, as permitted under the Copyright, Designs and Patents Act, 1988, no part of this publication may be reproduced, stored in a retrieval system, or transmitted in any form or by any means, electronic, electrical, chemical, mechanical, optical, photocopying, recording or otherwise, without prior written permission. All enquiries should be addressed to the publisher.

British Library Cataloguing in Publication Data

Burnham, Colin
   Beverly Hills Cars.
   1. American cars
   I. Title
   629.2220973

ISBN 1-85532-108-4

Photography and text Colin Burnham
Editor Nicholas Collins
Page design Angela Posen
Printed in Hong Kong

For my Dad, Jim Burnham

About the author

Colin Burnham is one of a rare breed of freelance photojournalists who writes to an equally high standard. His colourful pictures and words have appeared in numerous publications on both sides of the Atlantic, and his inimitable approach to American-related automotive subjects has been widely recognized through the Osprey Colour Series. This is his fifth compilation in recent years following *Air-cooled Volkswagens*, *Classic Volkswagens*, *California Classics* and *Bizarre Cars*. Though London-based, Burnham spends an increasing amount of time in Los Angeles focusing on all aspects of the car in Californian culture. He owns a 1971 Rover 'Carlton Club' Coupé.

All photographs taken on Kodachrome 64 using Olympus OM 35mm equipment

Front cover
Restaurateur Ciro Orsini owns this custom-built version of the classic late Thirties Cord 810. The car was built several years ago by a local enthusiast using mostly late-model GM parts and Orsini paid $68,000 for it. 'I park it outside the restaurant and people pull up to look and ask me what it is – it brings in dozens of customers'

Half-title page
Child portrait, Hollywood style

Title page
MR'N'MRS who? Mr & Mrs Rich'n'Famous of course!

Rear cover
The original 1955–57 'T-Bird' – as in the Beach Boys' 'We'll have fun fun fun till her daddy takes the T-Bird away' – captured the hearts of Americans in a special way and has since become an icon of popular culture. This is a 1957 example

For a catalogue of all books published by Osprey Automotive
please write to:
**The Marketing Department, Octopus Illustrated Books,
1st Floor, Michelin House, 81 Fulham Road, London SW3 6RB**

CONTENTS

INTRODUCTION	6
HI-ROLLERS	12
GOD BLESS AMERICA	24
LIMOUSINE DREAMS	48
NOUVEAU CLASSIQUE	62
ITALIAN STALLIONS	72
RICH MAN, PORSCHE MAN	82
MERCEDES MAGIC	94
MADE IN ENGLAND	106
AT YOUR SERVICE	116

Early Sixties Rolls-Royce Silver Cloud; just one of many classic cars cruising the streets of Beverly Hills . . .

INTRODUCTION

Automotive fantasies are great fun; they are never constrained by the petty limitations of real life. Money is no object; there are no insurance premiums, no petrol bills, no speeding fines. When it's time to dream, anything's possible. The following pages contain material that should whet the appetite of even the most fervent car buff. A selection of the world's finest automotive products built today, and yesterday, pictured in the wealthiest small city in the wealthiest nation on earth. Indeed, there's nowhere on earth quite like it.

Beverly Hills conjures up an adult Disneyland; a dazzling dream at the end of a rainbow where the streets are paved with gold (or, at least, gold-trimmed automobiles). Glamorous images created by movies and TV flood the imagination at the mere mention of its name. It would seem that everyone who lives in this sun-drenched utopia is a millionaire, with a millionaire's car and a millionaire's mansion. It's all super-sentimental and hyper-hedonistic compared to our everyday existence, and we console ourselves by thinking it's just not 'real life'. But, in fact, the reality comes close to the myth.

The city of Beverly Hills is situated about eight miles from the Pacific Ocean in the affluent Westside district of Los Angeles. Next door to Hollywood, the dream factory itself. For all its reputation, this 'city within a city' occupies less than six square miles; a mere speck on the map of Greater Los Angeles and adjoining counties. But in the same way as *Elay* is said to be detached from the rest of the continent, Beverly Hills is a virtual island within it; a world in cotton wool in which cultural diversity is measured not so much by race and class as by the number of millions one is worth. Ironically, this golden ghetto is little more than a stone's throw from the drab desolation of skid row, and just three hours drive from the Mexican border and some of the worst poverty known to humanity.

Unlike Disneyland, there's no entry fee into this Silver Cloud-cuckoo-land, and it is one of LA's tourist spots that lives up to all expectations. No ordinary working person would believe the opulence on display, both in the picture-perfect residential neighbourhoods and the world-famous shopping district. The regal excesses of this garden-city include the world's most exclusive stores in ritzy Rodeo Drive, lushly-planted, palm-lined avenues in which each and every home is fit for a king, and legendary hotels where prices start at *expensive*.

LA's Monaco-like haven (sister city of Cannes) has more Rolls-Royces, medical practitioners and gardeners per capita than any other place in the world, and personal trainers, shrinks and maids are considered *de rigeur* among many of its 35,000 residents. All commercial activity is of the white-collar, non-polluting kind, and everything is tightly regulated, even the dimensions of the signs. Crime is kept to the absolute minimum thanks to the city's well-staffed police department and dozens of private security guards, who patrol the streets day and night. And in stark contrast to most other parts of the 'Big Orange', it is a place where people actually *walk*.

Most of America's major urban centres had their boundaries rigidly set before the internal combustion engine belched to life. But Los Angeles was, and is, geographically amorphous, constantly evolving along an undisturbed, seemingly endless Pacific coastline. Thanks to Henry Ford and the assembly line process, the City of Angels quickly became the City of the Automobile; an irregular patchwork of suburban areas connected casually together under the vague concept of Greater Los Angeles. Like a volcanic eruption that solidifies into people and places but never quite stops moving, LA has become the world's largest

Left
Downtown Beverly Hills, looking eastwards over Little Santa Monica Blvd. towards Hollywood

urban sprawl; the quintessential 20th century *autopolis* in which distances are measured not in miles, but in estimated driving time. It is a city that was built on the driving ambition of immigrants from every corner of the US and the world, and it has been described as everything from 'forty-nine suburbs in search of a city' to 'Paradise with a lobotomy'. Clive James's recent observation – 'Los Angeles is so spread out that if you didn't own a car you could grow old and die trying to find the city limits' – is perhaps the most graphic yet. Certainly its ever-increasing population is perpetually on the move.

If California were to secede from the Union it would be one of the world's top 10 richest countries, and LA produces nearly half that economy. The money spent on sports cars alone exceeds the gross national product of many Third World nations. 'Success' is the bottom line and a constant preoccupation in this dollar-chasing society, and it seems that everyone has some kind of idea to sell. To stand any chance of achieving their dreams, *Angelenos* must develop an infallible level of self-esteem and project an image tailored to withstand ferocious competition. And, needless to say, he or she must also be seen to drive the 'right' car; it speaks volumes about your financial status and individual style (not to mention your inner betterment) in a place where outward appearances count for everything.

Since success breeds success, it is crucial, wheels-wise or otherwise, to appear as upwardly mobile as your budget will allow. Those who have made the big-time usually drive prestigious European marques, the very latest models. *Wannabes*, meanwhile, trade-up towards their dream machine at every opportunity, or customise their cars in some way in order to stand out from the crowd. Others prefer classic, quirky or altogether unique wheels; cars that serve both as transportation and artistic statement. Whatever the vehicle, it is a recognition of the driver's accomplishments, an extension of their personality, and pride of ownership is exhibited everywhere you look.

Automotive establishments abound and the weekly wash-and-wax takes on the sanctity of attendance at mass. Owners eye-up each other's cars while waiting at traffic lights or queuing at drive-thrus and relate to each other through personal statements on licence plates. Indeed nowhere is the innocence, the exuberance, the sheer indulgence of

America's long-running love affair with the automobile more celebrated. The downside of LA's sense of absolute mobility and freedom is that drivers spend an inordinate part of each working day behind the wheel on one of the nation's most congested freeway systems. To say nothing of the horrendous pollution problem...

Beverly Hills was once part of a Spanish land grant known as *El Rancho Rodeo de las Aguas*, or 'Ranch of the Circling Waters'. In 1852, two years after California became the 31st American state following a bloody battle with Mexico, Maria Rita Valdez sold the land for just over one dollar an acre. Around the turn of the century, the Amalgamated Oil Company, owned in part by the city's eventual founder, Burton E Green, bought out local beanfield owners in the hope of tapping the fat reservoir of wealth that Los Angeles first discovered in 1892. But only water was found. Thus the anxious investors re-named the oil company 'Rodeo Land and Water' and laid plans for a new city development, close-to but separate from the then-burgeoning City of Angels.

In 1906, two years before Ford unveiled the Model 'T', landscape architect Wilbur Cook designed the sinuously curving, tree-lined residential streets between Santa Monica and Sunset Boulevards, while the much-acclaimed Olmsted brothers created paved ways that wound up through the foothills of the Santa Monica mountain range as though in a landscape park. The thirty-block triangular-shaped commercial arena, now known as the 'Golden Triangle', followed later. The theme was distinctly up-market, and the opening of the grandiose Beverly Hills Hotel in 1912 did much to promote the scheme. By 1914, the wealthy had bought tracts and were moving in, and the city – named after President Taft's hideaway in Beverly Farms, Massachusetts – acquired municipal status. This enabled the original 550 registered voters to regulate the quality life in such a way that only the well-heeled could afford the entrance fee.

The city really became synonymous with opulence and glamour when Mary Pickford and Douglas Fairbanks set up home in the hills in 1920. The couple's lavish entertaining at their Pickfair estate attracted crowned heads and celebrities from all walks of life, and their glittering charity galas not only lent respectability to the fledgling movie industry, but to Beverly Hills as well. The new city developed rapidly and ever more exclusively as the Shiny Set followed in

A 'Presidential Suite' at the legendary 'Pink Palace' costs around $3000 per night, plus tax

the footsteps of the king and queen of the silent screen, and what was once a snake-infested expanse of sagebrush and tumbleweed became an enclave for the rich and famous. The apotheosis of the American Dream.

There are still many movie stars among the population. They – that is, America's equivalent to the Royal Family – live alongside the moguls who control the neighbouring entertainment industries and those who have made mega-bucks in all manner of commercial pursuits: real estate, fashion, fast food, cosmetic surgery, and so on. Often the kind of people who have made fast fortunes and can afford to pay cash for their mansions and gaudy trappings; the kind who might buy original art to match the living room décor and adorn the Mercedes with 24

carat gold trim. But it's not all celebrities and *nouveaux riche* types playing out the scenerio of the dream created by Hollywood. The city is also home to a strong Social Register society with old names that carry as much clout as Vanderbilt and Rockefeller in New York; the *cognoscenti* that have major boulevards named after them in LA. New money or old, loud or quiet, the residents of Beverly Hills nurture a strong sense of community that is both sophisticated and naive, exotic and banal. As Nancy Reagan said during her husband's term as Governor of California: 'Thank heavens we can escape to Beverly Hills on the weekends. No one in Sacramento can do hair' . . .

Most of the Old Money lives to the north of fabled Sunset Boulevard, up in the hills along with the *real* stars. They guard their privacy zealously, living in splendour behind daunting walls, with angry Dobermans and state-of-the-art security systems for reassurance. Down on the flats, however, between Santa Monica Boulevard and Sunset, New Money has elevated 'flaunting it' to an art form. This is where all the world's architectural styles come to co-habit and hatch the occasional hybrid, as if to mirror the fantasies of every ambitious immigrant in LA. The streets are an eclectic mix of Spanish haciendas, mock-Tudor mansions, Norman chateaux and neo-Classical temples, to name the most popular styles, and many homes have private tennis courts as well as the obligatory swimming pool and solarium. Lawns are like bowling greens, lollipop palms sway in the breeze, exotic vegetation flourishes in the Mediterranean climate; it's all too perfect to be true. And when Beverly Hills 90210 celebrates Christmas – eat your heart out Las Vegas!

Given the information so far, it is not surprising that the local shops cater only to those of the highest necessity. If you have to ask how much, you certainly can't afford it. Rodeo Drive's remarkable assemblage of boutiques, jewellery stores, art galleries and restaurants stand as testimony to the lavish, all-consuming lifestyle of LA's ritziest suburb, and most visitors are amazed. Some of the more exclusive stores trade by appointment only, and in many cases valet parking is provided – but then wouldn't you expect the personal touch if you were shopping for a gold-plated chess set or an original Rembrandt etching? It is a place where women from *Dallas* alight from stretch limos brandishing gold credit cards and jangling gold jewellery, and where tourists gawk in windows, gasp at the prices and take pictures of each other outside *Giorgio* to impress the folks back home. Street theatre *par excellence*.

In many of the city's ultra-chic restaurants, you cannot get a decent table unless you are some kind of celebrity, or at least look like one. At Café Rodeo, for example, the 'Beautiful People', with their open arms and oodles of air kisses, are always put on display at the front. Silk and gold-laden middle-aged matrons sit in the

The O'Neill Residence in North Rodeo Drive is a truly extravagant homage to Art Nouveau, built around 1980

inner sanctum, polishing their anecdotes and discussing the benefits of colon therapy and isolation tanks, while the 'Polyester People' will be sequestered in Siberia around the corner, safely off set. Each group has little in common other than being served by hopeful young actors and actresses waiting to be discovered.

Getting around in style has always been as important to the rich as having the right place to live in, and Beverly Hills has been playing cars all its life. Sitting alongside Rodeo Drive, you'll see more dream cars in the space of a few hours than you would at a major Motor Show. And all in motion.

Rolls-Royces are literally *everywhere*; a Corniche on every corner. Indeed, Beverly Hills's passion for the stately dreadnoughts borders on religious fervour. Seated on their hallowed leather upholstery, inhaling the ambrosia of consecrated walnut and fine oils, the privileged are transported onto a celestial plane. They also appreciate the quality and cachet associated with Jaguar, likewise that other prestigious English marque, Range Rover; an essential part of every gentleman sportsman's wardrobe. But it is the German car that massages most egos – that is, top-of-the-range Mercedes and Porsches (driving a run-of-the-mill BMW in this city makes you a second-class citizen). The really high-flyers whiz along in Ferraris, or other Italian supercars, all models, all colours.

Before the top-notch European imports changed the rules, every star worth his or her billing drove at least one Cadillac. According to LA folklore, at certain posh restaurants you could not get into the main parking lot unless you were in one. But nowadays it seems only those with memories of Hollywood's Golden Age buy American, since most contemporary Motown products are simply too brash for those who were educated to believe that European design meant *class*. Nevertheless, the all-encompassing nostalgia kick brings to light many *concours* Fifties convertibles at weekends alongside the finest examples of rare European sportscars. It's enough to give any car enthusiast automotive indigestion.

Car-crazy LA has seen a consistent trend to modify vehicles, no matter how good the original product may be. On the streets of Beverly Hills that can mean anything from subtle gold plating (*sic*) to a re-bodied Thirties-style roadster or an ultra-low quasi-racing Porsche. But that's not to say that every vehicle in the city and its environs is *desirable*. There are some truly dreadful blights on the asphalt landscape, each vying for attention in a place where nothing is ever too outrageous. The more eco-conscious or prudent citizens drive ubiquitous Japanese 'rice-burners', as they have done for years – though by comparison with many of the cars pictured, it has to be said, these lookalike econocars hold about as much appeal as a shopping trolley.

Beverly Hills Cars was shot in the weeks leading up to Christmas 1989, during which time the curtain rose on the Eastern Block, Zsa Zsa Gabor made the headlines after slapping a

The city streets represent an exciting and never-ending Motor Show

10

Beverly Hills cop, and the city concluded its 75th Diamond Jubilee celebrations. Leaving behind a damp, grey Britain, my companion and I followed in the footsteps of Raymond Chandler and his wife Cissy and made temporary home at laid-back Hermosa Beach – which proved to be the perfect antidote to a daily overdose of conspicuous consumption. We bought a VW convertible for anonymity (and what proved to be a negligable profit back home) and commuted daily along the six-lane 405 freeway; top down, smog-in-the-hair. The weather was exceptional (90 degrees on Christmas Day) and we experienced plenty of life-long memories along the way. Notably one Saturday.

Drifting along La Cienega Boulevard in West Hollywood, I spotted the unmistakable hairstyle of Rod Stewart jumping into a white Testarossa. 'Excellent', I thought. 'A major celeb' – just what the book needs'. One squealing *Starsky'n'Hutch* U-turn later, I was on his tail; weaving through the traffic with one hand, struggling to load a fresh roll of Kodachrome with the other. A genuine Hollywood car-chase, thankfully with no cops involved. Eventually we hit a red light. Out I jumped. A quick explanation, a nod from Rod, then click click click click. One superstar and supercar *in the can*. A definite cover shot, I felt sure.

Following this rush of excitement, I pulled over to catch my breath. We were now on the hippest section of Sunset Strip, feeling suitably chuffed. Glancing across the street at a group of bikers, another famous coiff' caught my eye: 'Hey that's Billy Idol!' And sure enough it was. The British Rock'n'Roll hero was sitting astride his Harley chopper outside a trendy 'straunt looking exactly like a Rock'n'Roll hero should; *The Wild One* thirty-five years on, with spiky peroxide hair. Casting my British reserve aside, I switched into maximum street-wise mode and approached the charismatic figure. Idol was surrounded by a somewhat intimidating bunch of leather-clad *doods* and the closer I got the more I could relate to the reticent character of Woody Allen in *Play it Again Sam*. Using my faithful old 35-mill to gain his attention, we exchanged a few short words and I took as many candid pictures as was possible under the circumstances. He was used to it.

Walking back to the car, gleeful at the thought of a fat cheque arriving from one of the tabloid Sunday supplements back home, I couldn't believe my luck. I'd suddenly become a member of the *paparazzi* and it wasn't such a bad job. I slipped on the telephoto and took some more pictures as the rock star rode away, *just for the album* as they say.

There were four frames left on the roll but car-spotting was now off the menu; this film was going straight to the lab. Aiming the camera at the nearest point of interest, I pressed the power-winder: click click click click click click click – a momentary pause – click click click click. The adrenalin began pumping again – this time for the wrong reasons. The film counter was at 'E' (end of the roll) but the winder continued to operate. I knew what had happened straight away but my 'Positive Mental Attitude' refused to accept it: the film leader had not engaged properly in the spool so the film was completely unexposed. As the realization of my blunder sank in, the sky turned a darker shade of blue. I'd blown it, and there would be no second chance.

So much for show-business...

Colin Burnham
London, England, 1991

HI-ROLLERS

The only thing one needs to know about a Rolls-Royce is that it's a Rolls-Royce. It doesn't really matter whether it's quick or slow, good value or robbery, state-of-the-art or archaic; it's a *Rolls*, synonymous with success, perceived as the best. And with a Rolls-Royce in ritzy Rodeo Drive, you really have 'arrived' — probably with someone else to do the driving...

Perfectly restored Rolls-Royce Silver Cloud III, pictured outside Giorgio Armani's West Coast flagship store in Rodeo Drive. This majestic motor car cost £4660 when new in 1965, roughly half the price of an Armani ladies evening jacket twenty five years later

Right
Status symbol: The Silver Lady, or Spirit of Ecstasy, was designed in 1911 by Charles Sykes, an eminent sculptor, whose objective was 'to depict a woman who has selected road travel as her supreme delight and has alighted on the prow of a Rolls-Royce motor car to travel in the freshness of air and the musical sound of her fluttering draperies'

Below
In contrast to the UK, where a 'cherished' number plate is looked upon as a symbol of affluence, licensing laws in the USA allow freedom of expression for every car owner in return for a minimal fee. The individual's statement is limited by a maximum number of letters and/or digits, and for many it is a contest to see who can come up with the most complex, arcane or esoteric message ('bumper stumper'). Others prefer to state the obvious

Left
For Her: A black Camargue

Below
For Him: A gold Silver Spur

Previous page & above
'I was so impressed, I bought the company!': Victor Kiam's well-known catch-phrase could easily be that of John Wick (pictured). This wealthy American businessman was, indeed, so impressed with the Hooper-built Rolls-Royce he had bought for his own personal use, that within a short time he acquired the long-established London coachbuilding company lock, stock and barrel. John's pride and joy is a unique Silver Spirit, converted from a four-door into a two-door and featuring a smaller limousine-style rear window, wire wheels, colour-coded bumpers, and so on. The licence plate reveals where John spends most of his time, but he always takes the car with him when business calls in Los Angeles. What better place, likewise what better light in which to promote the company?

The hard commercial world seems so much further away thanks to the combination of classical music played through a multi-speaker sound system, comfortable leather seating, superb air conditioning and, of course, the smoothest ride imaginable

Elegant Silver Cloud S.III coupé is an early Sixties product of Mulliner Park Ward, the 'other' London-based coachbuilders affiliated to Rolls-Royce. Its eye-catching paint job, smoked glass and whitewall tyres are typical 'Elay' accoutrements

John Turton, a former British schoolboy discus champion, is chauffeur to a multi-millionaire who wheels and deals in a stretched Rolls-Royce Silver Spur II. John has been working in the Golden State since 1975, the year in which he was sent out on a short-term track scholarship and decided to stay. As an illegal alien, the Berkshire boy had numerous casual jobs before finding a reasonably profitable niche as a chauffeur-cum-bodyguard, but after 15 years he is considering returning to Britain with his American wife and young son. 'You get fed up with all the glitter and glamour out here,' sighs John. 'And the schools – they're full of drug dealers.'

D H Lawrence, the English novelist who died in 1930, wrote: 'Los Angeles is silly – much motoring, me rather tired and vague with it. California is a queer place – in a way, it has turned its back on the world and looks into the void Pacific. It is absolutely selfish, very empty, but not false, and at least, not full of false effort. I don't want to live here, but a stay rather amuses me. It's a sort of crazy sensible.' One can only assume the owner of the Rolls has similar sentiments

'Budget' in Beverly Hills bears no relation to 'budget' anywhere else. Here, for a mere $500 a day, you can rent the Corniche convertible driven by Joan Collins in Dynasty, or the red Ferrari piloted by Eddie Murphy in Beverly Hills Cop II. Or any one of numerous speciality/exotic cars that may be available. According to Corky Rice, co-owner of the franchise, customers usually fall into one of two distinct categories: 'The moneyed lifestyle maintainer' whose Jaguar XJS is being repaired and rents an identical model for continuity's sake; 'Mr or Ms Flash' – with a valid credit card and little else – who is out to impress, the cost be damned. (A nurse who rented a Lamborghini Countach for six weekends is a case in point. The Budget staff guessed she wanted to meet somebody as well off as she pretended to be)

GOD BLESS AMERICA

The days of the gas-guzzler may be numbered, but the traditions of Motown will remain alive for some time yet in Beverly Hills. 'Yank tanks' of all ages – from the occasional pre-war sedan, Fifties convertible and Sixties coupé, through all the later prestige models – can be seen cruising the streets in all their brash splendour. More often than not, they are driven by those who can remember the days when Los Angeles freeways were rarely congested, when gas was cheap, too much chrome was just enough and the dynamo hum of a heavy-breathing V8 was music to the ears. While Detroit continues to cater for the traditional American taste in cars, naturally, this affluent market segment is loath to give them up...

Located on San Vicente Boulevard (opposite the Hard Rock Café), Eddie Blake's Tail O'the Pup hot dog stand is a cultural landmark in West LA. The 17 ft long stucco hot dog, complete with yellow mustard, was constructed in 1946 and is one of the best remaining examples of roadside pop architecture. The 'dogs are great, too

Often billed as the fastest sports car ever produced, the AC Cobra was an unusual Anglo-American hybrid. The bodies and chassis were shipped complete from AC's factory in Thames Ditton to Carroll Shelby's workshops in Los Angeles where a Ford V8 – first a 260 cu in version, then a 289 and ultimately an awesome 427 big-block – was dropped in. Fewer than 350 of these aluminium-bodied missiles were built between 1962–1967, and the Cobra has since become the most imitated car in the replica circus

The Sharper Image is a trendy establishment selling the kind of 'designer' accessories that go hand-in-hand with the affluent California lifestyle. There you will find such devices as the 'Alcomax Breath Analyzer', the 'Auto Pet Food Server' and the de rigeur 'Telephone Tap Detector'. What's more, in the weeks leading up to Christmas 1989, the store also offered this flawless, 12,000 miles-from-new 1956 Chevrolet Corvette. The asking price? $75,000

The Beverly Center, just east of Beverly Hills, stretches a full traffic light-to-traffic light block on each of its four sides. Opened in 1982 and comprising some 200 shops and restaurants and 14 cinemas, critics have described Tinseltown's largest retail building as an 'affront to architecture'. But it does have one interesting exterior feature: a 1959 Cadillac plunging through the roof of the Hard Rock Café

Super-trendy 1959 Cadillac settee, courtesy of Larry Parker's Diner, provides the perfect resting place for poseurs on South Beverly Drive 24 hours a day. 'Lumiere' is the name of the lady who sat down as this picture was being set up, and it is to be noted she is an actress, composer, model and a recording artist currently assigned to Atlantic Records . . .

Left
Tail-fin, 1960 Ford Sunliner. A reminder of a more innocently optimistic period when Americans bought 'American' and the vast majority of Teutonic imports were simply called 'Bugs' . . .

Above
Beverly Hills circa 1960? It could be. In profile is a '60 Buick Electra 225 (its length in inches), in the drive the infamous late Fifties Ford Edsel. Both behemoths are considered classic 'land yachts' thirty years on

Next time you're negotiating with the seller of a classic such as this 1964 Lincoln Continental convertible, you could try quoting the late A P Giannini, founder of Bank of America. 'No man actually owns a fortune,' he said. 'It owns him'

Below
The bumper sticker continues to be one of the favourite forms of expression on the streets of LA. According to a survey in the Los Angeles Times, 18 per cent of car owners in the SoCal metropolis display at least one message on the back of their vehicles. The owner of this 1964 Ford makes up for the rest

Overleaf
Swapping the manufacturer's grille for a custom-made 'tooth' effect item was a popular trick among car customizers back in the Fifties, but they were never like this! 'Jaws' is the rather exceptional handiwork of Michal Serebrini, a sculptor-cum-jewellery maker-cum-exhibitionist

For struggling actor, Dennis Woodruff, the road to fame and fortune is travelled in the 'Woodruffmobile', the most bizarre Lincoln Continental. The car is literally covered in self-promotional material in the hope that its owner will be discovered and become, in his own words, 'a legend, like James Dean'. Woodruff's blatant campaign has also included hundreds of posters, T-shirts and thrift shop ties emblazoned with his name, and all his efforts have not been in vain (sic). He has appeared, albeit briefly, in Star Trek, Quantum Leap and a Phil Collins pop video, and his name is even being used as a synonym ('Don't be such a Woodruff!') in certain Hollywood bars. To which he reacts: 'Do be a Woodruff. Woodruff is a winner. Woodruff is a guy that, through living his life and finding himself, realizes his purpose is to be really important to people – though I don't want to take myself too seriously . . .'

Deals on wheels: LA's cellular phone system went on-line in 1984, when transforming your motor into a mobile office was less about making a commitment to efficiency and more about making a status gesture. Now, most onlookers are far less impressed. Young entrepreneurs like Bill Asher (pictured) – son of actress Elizabeth Montgomery, who played Samantha in the TV series Bewitched *– no longer make calls simply to say, 'Hey, guess where I'm calling from?'*

Long-time Beverly Hills resident, Pam Morel, and her all-white 1976 Pontiac Grand Prix. 1976 marked Pontiac's 50th anniversary as a producer of quality cars, and Pam's nigh-on 18 ft long, 2-ton coupe has all the special features one would expect from a top-of-the-range model. 'I took delivery in February '76 and have loved it ever since,' says Pam. 'But I'll get something smaller next time. Trouble is, cars today – they're made of spit!'

Simulated wire wheel, mounted for all the world to see, on bustle-back Cadillac Seville Cabriolet; a popular Cad' with the lady motorists

Season's Greetings: 1952 Cadillac, as seen on Christmas Day, 1989

Left
The Beverly Wilshire is the grand old hotel of downtown Beverly Hills, first opened in 1928. Its guest list includes kings, presidents and movie stars – Axel Foley (Eddie Murphy) stayed there in Beverly Hills Cop *– and opulence is most certainly the watchword. Located on Wilshire Boulevard at the foot of Rodeo Drive, the hotel puts on this impressive show of lights during the festive season*

Below
Let there be lights! Many residents put on the most spectacular lighting displays at Christmas in an effort to 'out-do thy neighbour'. Did someone mention energy conservation?

43

Above
Love me, love my dog

Right
Sales of four-wheel drive vehicles escalated beyond all proportion during the 1980s. This image – driver posing in immaculate Jeep on 'Mainstreet' – tells why

'Chrome to the bone!' Not so long ago, the words 'Harley-Davidson' conjured up an image of rampaging Hell's Angels; social outlaws who kicked the system and anyone who stood in their way. In 1990, however, those same 'hogs' are straddled by doctors, lawyers and accountants in search of a low-calorie brunch. These 'Rubbies' (Rich Urban Bikers) have followed Hollywood heavies like Sylvester Stallone, Mickey Rourke and Patrick Swayze to be the new bad boys on big bikes – at least on weekends

Girl on a motorcycle: According to official 1989 Harley-Davidson statistics, Harley riders' median annual income is over $36,000; 56 per cent are married; 38 per cent attended college; 7 per cent are women

LIMOUSINE DREAMS

The sheer physical presence of a 'stretch limo' makes it the most conspicuous car on the road at any given moment. Be it gliding along Sunset Boulevard *en route* to a Hollywood function or double-parked outside an exclusive boutique in Rodeo Drive, this ultimate symbol of American excess just begs to be noticed. Those which are not owned by individuals who wish to maintain a certain profile are operated by small hire companies who cater to large egos and/or expense accounts; people who need to impress other people for whatever reason. The vehicles are usually dealer-supplied Cadillacs or Lincolns, at least twice the length of the average car and equipped with TV, video, phone, fax, cocktail cabinet and, of course, a suitably-attired chauffeur. But not always. One coachbuilding concern in LA specializes in stretching some of the most unlikely vehicles to ridiculous lengths, including the minuscule Fiat 500 and, would you believe, a Lamborghini Countach...

Stretched Lincoln Town Car – its occupants bound for a glittering function, no doubt...

'Cadillac... universal symbol of achievement.' Or so it was claimed in the company's 1959 advertising campaign. Nowadays, the Cadillac emblem has considerably less cachet in car-mad Southern California, though it is still favoured by many limo drivers and those with memories of LA's swanky past...

The wink of chrome on a limousine as the driving team rush off to collect another important client

To stretch or not to stretch? That is the question many coachbuilders will ask themselves if/when the trend becomes passé. Carrol & Co, meanwhile, will no doubt continue to dress nice, conservative Beverly Hills gentlemen as they have done for 40 years

Above
'Before' and 'after' Lincoln Town Cars

Overleaf
Limousine companies go to extraordinary lengths to cater for those who wish to impress the rest of us

Above
The 'Jag-wahr' for fat cats, a stretched XJ 'sedan' with 'tints'

Right
Pictured at the intersection of Santa Monica Boulevard and Bedford Drive is the classic Rolls-Royce Phantom VI limousine. Twenty-feet long, the seven-seater Phantom model dates back to 1959 and its list of passengers includes the British Royal Family, numerous heads of state, and VIPs the world over. This one serves as a courtesy car to guests at the Beverly Hilton. (The lushness of Beverly Hills – there are as many trees as there are residents – belies the fact that LA is built on a desert)

Overleaf
Show cars aside, this must be one of the longest working limos in LA. It was standing-by at a movie location; the star's car, of course

Left
It has been calculated that the average American motorist spends a full six months of his/her life waiting at red traffic lights. An astonishing statistic, but one that anyone who commutes along the Beverly Hills section of Santa Monica Boulevard during the rush hours would certainly believe. Traffic moves at an escargot's pace in both directions; in this case eastwards, towards Hollywood and the San Gabriel Mountains

Above
Reflection of one of LA's 128 assorted varieties of palm trees – or could it be the elusive Rod Stewart?

NOUVEAU CLASSIQUE

Back in the late Twenties and early Thirties, when Beverly Hills was earning its reputation as a haven for the rich and famous, car companies like Duesenberg, Auburn and Packhard hand-crafted the most fantastic automobiles for the fortunate minority. With curvaceous wings, mile-long bonnets, chrome-plated vertical grilles, free-standing headlamps, wire wheels and flexible external exhaust pipes, these and other great marques are now regarded as the all-time classics, with price tags to suit. But for those enamoured of such styling fillips and the glamorous age they represent today, there are 'neo-classics': fibreglass-bodied 1930s-style roadsters, custom-built around contemporary Detroit components. To the layman they look like the real thing and therein lies most of their appeal. *Aficionados*, however, generally agree that the majority of these offerings should not be taken too seriously...

The Heritage is an impressive reproduction of the classic 1934 Mercedes 500K, albeit based on Chevrolet Camaro running gear

Previous page
The V8-engined Clenet is an exceptional example of the neo-classic style. Genuinely well built with a price tag to suit, the model is now out of production but much sought-after by collectors. This one is owned by the proprietor of Barakat, *an emporium of Old World antiquities in Rodeo Drive*

Left
The Clenet's elaborate exhaust system conjures up those glamorous Hollywood images of the 1930s

The somewhat baroque Excalibur Phaeton evolved in the early Sixties and was produced until late 1990, when the Milwaukee manufacturer went down the tubes. This is a Seventies model featuring a 300 bhp Chevy V8, Corvette independent suspension and leather-lined interior. (Cute name for a drug store in Beverly Hills, eh)

Quintessential Beverly Hills car or crass concoction of fibreglass and chrome? Based around the cabin section of a Ford Mustang, the Zimmer Golden Spirit embodies all the unashamed opulence associated with the city as well as all the styling prerequisites of a true 'nouveau classique' – and then some. The ultimate display of flash and cash, this model is said to be a favourite among drug barons, pimps and mobsters...

Star's car? This pseudo-traditional roadster was spotted in the car park at Le Dôme on Sunset Boulevard, a French restaurant frequented by some of the biggest names in Hollywood

Clenet roadster – as in Alain Clenet of Clenet Coachworks, Santa Barbara – may look like a regular retro kit car to some, but it's something of a milestone in neo-classic circles

ITALIAN STALLIONS

After eight decades of unquestioned sovereignty, the car has at last been reduced from mythic proportions to mere transportation – a necessary evil, even – in terms of world consciousness. Its place in our dreams of glory, however, has not diminished. Somewhere, deep within, we still believe in the car as Phaeton, a marvellous chariot that can give us the speed of the wind and make us gods three times over. As we squeeze dutifully into efficient little machines with humble, domesticated names like Uno, Metro and Civic, though we may know we are doing the responsible thing, there is a void in our souls. Enter the supersonic sculptures from Italy, seemingly from another planet. Enter the 'exoticars' with evocative names like Lamborghini Countach, Ferrari Testarossa, Maserati Merak. Enter a fantasy in which you're flying down an empty highway at 180 mph...

Probably in the music biz or advertising; this is the man who drives a fire-engine red Ferrari...

Ferrari 308GTS *could be the car from* Magnum P.I., *but, like Don Johnson and his* Miami Vice *Testarossa (overleaf), Tom Selleck was nowhere to be seen . . .*

Above
Angelenos and their cars are like cowboys and their horses — almost inseparable. To own a prancing horse, a Ferrari, in LA is to be a modern-day Roy Rogers. Ferrari aficionados generally consider the Mondial model the poor relation of the stable, but to most Angelenos it's a Ferrari, period. An object of desire

Left
Other notable streets on the map include Brilliant Drive, Utopia Avenue and Rich Street, to name but a few

Above
Glamorous 'Redhead': The name 'Testarossa' is derived from the red rocker covers on the sports-racing Ferraris of the 1950s. This beauty was spotted opposite Ed Debevic's, West LA's leading neo-Fifties diner

Right
A rare sight, even in Rodeo Drive, and both illegally parked. The combined value of these Testarossas is close to half-a-million dollars – would you worry about a $13 citation?

Street scene, Sunset Boulevard

Left
The Midas Touch: If all that glitters is not gold, then De Tomaso Pantera owner, George Peloquin, has wasted a lot of time and money on a fool's mission. PUR GOLD (as the licence plate reads) is a mega-buck motor that simply begs to be noticed and its 24-carat driveline defies description. Only in The Golden State . . .

Above
Still built by DeTomaso in Modena, Italy and once sold through Ford dealerships in the USA, the Italo-American Pantera is a sleek combination of sexy styling and V8 reliability. Moreover, this blackberry-coloured 1974 model benefits from the full California beauty treatment: flared arches, moulded front spoiler, filled body seams, additional air intakes, Center Line wheels and, of course, gold-plated trim. Wow!

The outrageous 180 mph Lamborghini Countach may be the stuff of myth and magic, a 'Rambo on wheels' even, but it did not impress this stern LA parking officer. In fact, he took great pleasure in slapping a ticket on it!

If the difference between men and boys is the price of their toys, Select Motor Cars on Sunset Boulevard must be the ultimate adult toy shop. It's a stone's throw from the Beverly Hills/Hollywood border

RICH MAN, PORSCHE MAN

The current generation of high-level American consumers, having ripened in an atmosphere of prosperity and unprecedented exposure to television advertising, generally subscribes to the belief that European craftsmanship and design are superior to their American counterparts. Certainly that is the case within the Southern California car market. Many Angelenos aspire wholeheartedly to ownership of Teutonic imports, and for those living life in the fast lane that means a brand new Porsche; the ultimate driving experience if you can afford it. Needless to say, Porsches are as common a sight in West LA as Peugeots are in Paris – well, almost...

Head-turner: Porsche 928 with mirror-like rims carries plenty of status on the streets of LA. Ditto any Porsche product (see overleaf)

Left
Every field of endeavour, no matter how frivolous, has acceptable terminology that, if used properly, separates the aficionados from the poseurs. For example, in the automotive world, not knowing how to pronounce the name Porsche ('Porsher') can send one plunging through the social orders like a piano down an elevator shaft. Vanity plates reveal even more about where a person is 'at'

Above
There's something very lovable about the curvy Fifties Porsche Speedster. Its shape is one that has become a statement of good design for its time, accentuated by the low-cut windscreen and the bunker-effect ragtop. Many have survived the ravages of time in California's benevolent climate, but they are gradually being snapped-up by European investors. This one's a repro'

Right
Traditionally, half the Porsches made in Stuttgart are exported to the USA and half of those are sold in California. Thus the City of Angels has long been a hot-bed of styling ideas for the 911. Flat-nose Porkers proliferate in the better parts of town

Below
Porsches galore at Zipper Porsche, Wilshire Boulevard

Opposite
Road & Track once described the Porsche 911 Turbo as 'Outrageous, simply outrageous. How else can we describe a car with this much acceleration and unrestrained performance?' Amen

Left
Californian motorists, generally speaking, pay much attention to the wheels on their cars; original rims are often swapped for aftermarket items or otherwise embellished. But there's nothing quite like a Porsche 'cookie-cutter' with highly-polished spokes

Right
Ferdinand Porsche would surely approve: Bible black Beetle with lowered suspension, Porsche 914 rims and, most probably, a double-the-standard-horsepower motor is a typical example of the now-ubiquitous 'Cal-look' Volkswagen. The look evolved in Orange County, south of Los Angeles, in the early Seventies and has spawned a multi-million dollar VW accessory industry

Below right
Shades of the Sixties. According to its creator, Leo (left), this bizarre Bug represents 'action without thought'. Enough said

The late-afternoon sun spotlights a Porsche 911 Turbo as it whirrs past the most palatial, terracotta-topped residence in palm-lined Foothill Drive

MERCEDES MAGIC

Low profile and high quality are the traditional Mercedes-Benz hallmarks, with endless durability and strong residual value as icing on the cake. The Mercedes, any *standard* Mercedes that is, manages to be stylish yet anonymous, with no extrovert driver appeal but an isidious way of sneaking into the owner's affections (the aftermarket industry capitalizes on this fact). This prestigious German marque practically rules the lucrative 'executive' sector of the market thanks to the inherent hewn-from-solid quality and faultless attention to detail. When you work hard for your pleasures, the Mercedes aura can make you feel very well rewarded...

Nose job: A 190 SL Roadster shows its most attractive feature behind that much-maligned American artefact, the parking meter (invented by one Carl C Magee and first installed in Oklahoma City in 1935)

Above
The four-door Type 300 Cabriolet was a car for well-heeled Germans of the Fifties. Perhaps this one accompanied its original owner to the USA?

Right
Scene stealer: Mercedes-Benz built less than two thousand 300 SL Roadsters between 1957–1963, but only in Beverly Hills 90210 would you see one left unattended on the street. Most have ended-up under dust covers in temperature-controlled garages, left to appreciate rather than be appreciated. True spirits like Dr Norman Leaf, the plastic surgeon who owns this '58, drive and enjoy

Monochrome exterior, mirror-like rims, mega-watt sound system – this ground-hugging Merc' exemplifies contemporary customising trends

'Sports Lightweight' speeding westwards along Sunset Boulevard towards Bel-Air

Los Angeles staged the Summer Olympics in 1984, and the DMV (Department of Motor Vehicles) celebrated the occasion by offering special commemorative licence plates to those wishing to make a statement. This one's pretty succinct

If you thought a driveway was something you drove up to get home — forget it. Excess, not access, is the name of the game in Beverly Hills . . .

Right
More mirror-effect rims, more leopard skin...

Below
Any colour, so long as it's yellow! The man behind *Fred Hayman. Beverly Hills,* hands out gifts to his parking staff on Christmas Eve as they attend to his eye-catching 450 SL. Hayman's landmark store (formerly Giorgio) was the basis of Judith Krantz's novel and later television series Scruples, which was about people who didn't have any. But why the obsession with yellow? 'Yellow is California,' explains Fred. 'It's Beverly Hills, it's happy. Yellow is simply what we are'

Incongruous image: Like New York, the City of Angels is a major urban centre for American Sikhs. The religion eschews the use of tobacco, coffee, alcohol and meat, and the cutting of any body hair is also forbidden. This member of the sect, of which there are an estimated 5,000 in LA, was striding towards a gathering in a posh Beverly Hills restaurant

Standing on the corner watching all the Mercs go by . . .

Thanks to Beverly Hills Cop *and other movies, the rusty metal signs that mark the city's boundaries are familiar the world over*

MADE IN ENGLAND

Exclusive Beverly Hills has more than its fair share of most things in life, including Anglophiles; those who long to age their new money in a gentlemanly or lady-like way and admire the British sense of understatement, assurance and disdain. Or rather, those with a penchant for the trappings of the English upper-crust. They live in butler-and-Bolinger mock-Tudor mansions, they dress in a tweedy kind of elegance and, in most cases, they drive a *Jag-wahr*; the marque of the true connoisseur (according to most advertisements). In Tinseltown, if you truly believe you are as educated and civilized as those English dudes and if you drive one of their cars, then by golly, you are . . .

Prior to the explosion of interest in classic cars in the Eighties, Jaguar's big Fifties' saloons were generally thought of as wallowing whales that rotted almost as fast as they guzzled gas. Now, most surviving examples on either side of the 'sink' are restored to their former glory and serve as testament to their owners' discerning taste. Witness this Mk IX, resplendant with its two-tone finish and whitewalls

107

Left

'It matches my personality – it's spunky,' says Arlene Sand of her 1967 Austin Healey 3000 Mk III. The self-confessed Anglophile, pictured on Wilshire Boulevard after a shopping spree at Neiman-Marcus, bought the classic British Racing Green sportscar from its original owner in 1987. It has been meticulously maintained, never raced or rallied, and rarely driven in the rain. Naturally, Arlene enjoys all the compliments inspired by her pride and joy though she is quick to bemoan the cash offers that sometimes follow. 'I know what they fetch over in England,' says Arlene. 'But this one's staying right here with me.'

Above

'The V-12, 263-Horsepower XJ-S Convertible. For Those Few Upon Whom The Sun Always Shines.' (Headline, 1991 Jaguar advertisement)

Long, graceful, lithe and unmistakably feline, the Jaguar E-type or 'XKE' is surely one of the most beautiful cars ever designed – certainly one of the most desirable. With qualities to match any Italian exoticar, but at a fraction of the cost, the 150mph projectile was in a class of its own for most of its 14-year production run. Of the 70,000 made between 1961 and 1975 roughly 55,000 were left-hand drive, though many have since been shipped back to Blighty and converted to RHD. Cool 'plate for an ex-pat, eh

Fantasy architecture flourished in the US in the Thirties and Forties culminating in an entire kingdom of make-believe, Disneyland, in the mid-Fifties. Spadena House, Henry Oliver's quintessential witch's house, was originally built as a movie set for a production company in Culver City in 1921 and later moved to its present location in the heart of Beverly Hills, where it stands as a monument to California kitsch

Left
One of the many imposing billboards between Beverly Hills and the nearby San Diego Freeway 405. Needless to say, such apparitions cannot be erected within the city limits

Below
With its swooping cantilevered concrete canopy, Beverly Hills' Union 76 gas station is a classic example of Fifties American commercial architecture. It was designed originally for Los Angeles International AIrport (where it was to complement the famous flying saucer-like Theme Building), but was erected at the corner of Little Santa Monica Boulevard and Crescent Drive in 1965

You are what you drive: Americans bought about 80 per cent of the 100,000 MGAs made between 1955–1962, and now, like the majority of cars conceived in the Fifties, the model has undeniable street chic

AT YOUR SERVICE

While most Beverly Hills cars are a source of pleasure for their owners, some have to work hard for a living. These are the humble servants of the community; vehicles that reflect their driver's status in a more explicit way. They include thrice-round-the-clock taxis wallowing back and forth to Los Angeles airport; wheezing buses transporting household employees back to the other side of the tracks; 'Black'n'white' cop cars patrolling the streets twenty-four hours a day. What the 'other half' drive, you might say...

'Too bad all the people who know how to run the country are busy driving cabs and cutting hair.'
George Burns

Beverly Hills was once just a whistle-stop between Los Angeles and the Pacific Ocean, known as Morocco Junction. It was not until the Beverly Hills Hotel was built in 1912 that it began to attract attention. The Pacific Electric Company, seeing a need to transport visitors to the hotel, laid tracks in the centre of Rodeo Drive for a tram which is now the four-wheeled Beverly Hills Trolley. This unique vehicle tours the city every half-hour and the ride is free!

Despite the valiant efforts of the Southern California RTD (Rapid Transit District), travelling anywhere in Los Angeles by bus can be very unpleasant and time-consuming. This rudimentary public transport system, in effect, serves only those who cannot afford a car

Left
Street vendor at work, Wilshire Boulevard. Every third person in LA County is Hispanic, and Hispanic children make up more than half the student body in city schools. An estimated three million have fled the poverty of Mexico and the strife of Central America to the 'Big Orange'. But for many, the promise of Los Angeles outstrips the reality. **Undocumentados** *often work at less than the minimum wage ($4.35 an hour in 1990), yet those who staff kitchens and sweatshops count themselves well-off compared with the hundreds who attempt to sell fruit and nuts at traffic lights*

Below
Gardener seeks even break

Left
Rodeo Cab's *star driver, that's Ida Lee, a 70 year-old grandmother and part-time actress. Ida became a cab driver in 1984, soon after she arrived in Los Angeles hoping to land some mature parts in TV and movie productions: 'I love to drive, and by the time I'd spent six months auditioning all over LA, I knew my way around,' said the former cable TV talk-show host, who, at 49, walked out of a faltering marriage and went back to college. As the most senior of only a handful of female cabbies in LA, Ida has been robbed, cheated, flirted with, crashed into, and even ostracized by her male colleagues. As an actress, she has appeared in such movies as* The Right Stuff *and* The Fabulous Baker Boys, *and was the star of* Grandmother's House. *This amazing senior citizen says she intends 'to keep right on doing things that make me enthusiastic about life'*

Right
Keeping up appearances: Cab drivers have a somewhat unscrupulous reputation in Los Angeles, but Rodeo Cab *drivers, like other service employees in Beverly Hills, are kept on tighter reins than most*

Below left
LA Taxi *is one of the more established cab companies in the metropolis. This one also serves as a rolling billboard for one of LA's two National Football League teams: the LA Raiders, or 'black-shirts'*

Below right
Beverly Hills Cab, *as seen through one of the many thousands of new boxes owned by* The Los Angeles Times. *The date is December 1, 1989*

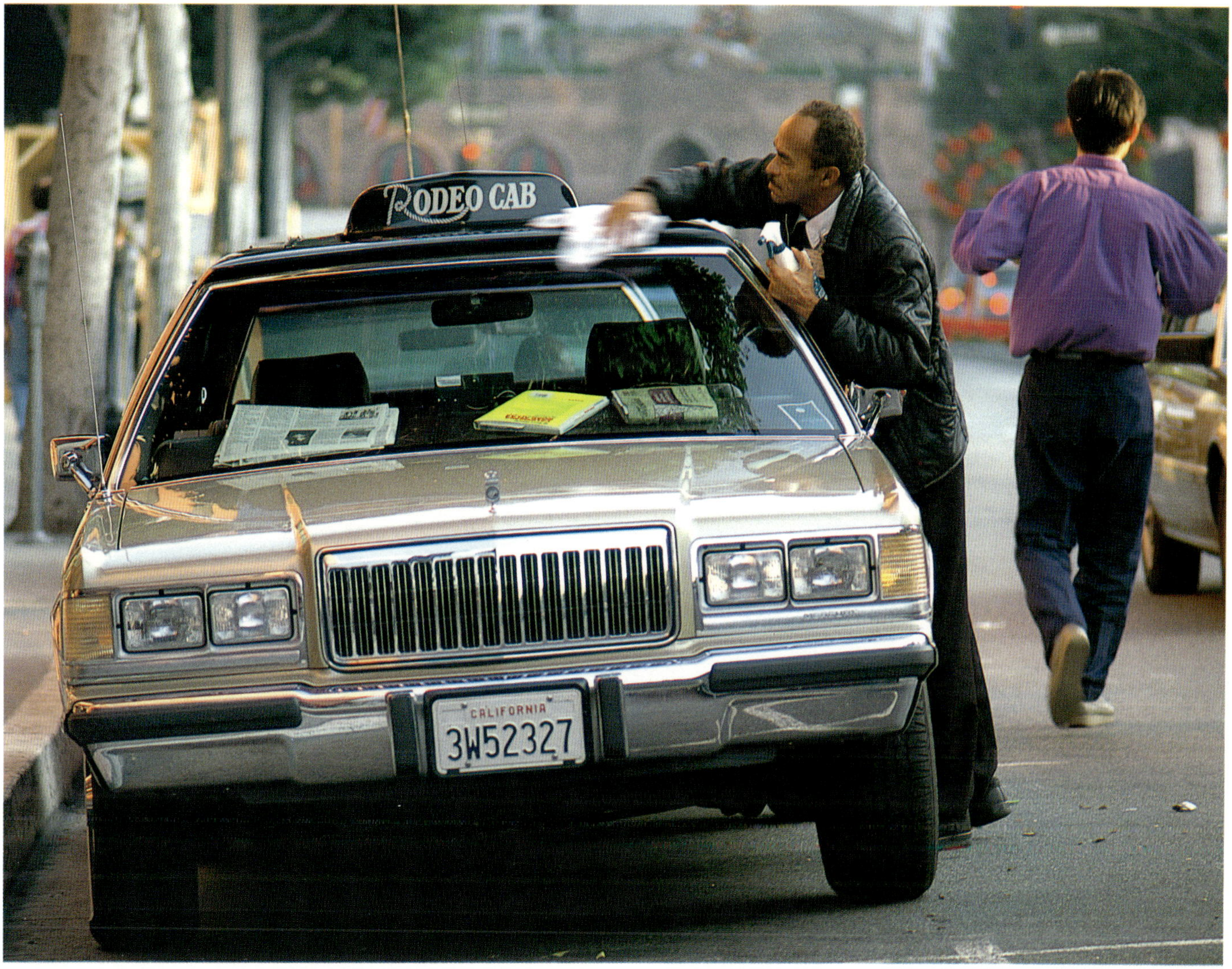
RODEO CAB
CALIFORNIA
3W52327

RAIDERS
COMMITMENT TO EXCELLENCE
4409
LA taxi

BEVERLY HILLS CAB
Los Angeles Times
s Angeles Tim

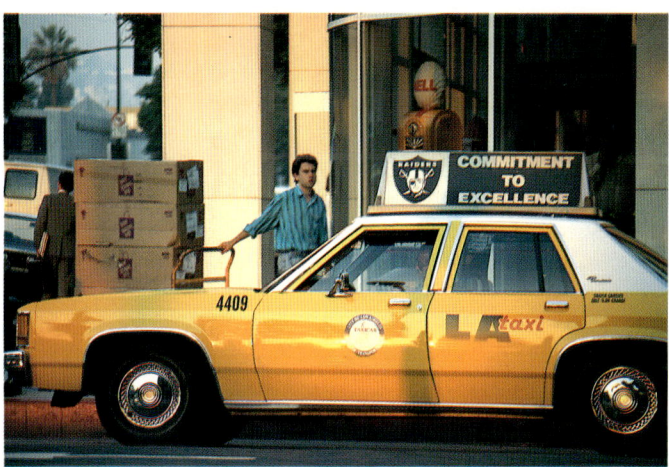

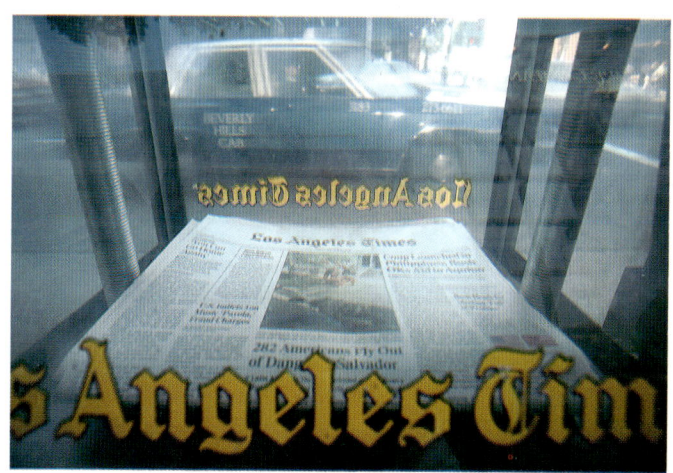

Above
Down and Out in Beverly Hills: Undesirables are quickly escorted beyond the city limits or otherwise detained in one of LA's 10,000 temporary cells. Beverly Hills cops have long been renowned for their polite but firm xenophobia

Right
The ubiquitous Beverly Hills Police car is never far away in the city that, in 1940, Raymond Chandler described as 'the best policed four square miles in California'

The Fire Department takes exceptional pride in its machinery, as do the other emergency services. They are all reached by dialling '911'

This nippy little contraption delivers umpteen parking citations per hour

What better advertising for a shoe shop? The 'Bootmobile' is based on a VW Beetle chassis and is driven daily to and from Quality Shoe Repair *on Santa Monica Boulevard, a few miles west of Beverly Hills city limit*